male

2.75–4"

2.5–4.2"

Pipevine Swallowtail
black overall; males have iridescent greenish-blue hindwings; fast, low flight; distasteful to many predators

Black Swallowtail
black with yellow bands of spots; black-centered eyespot near tail; common garden butterfly

female

male

3.5–5.5"

3.5–5"

Eastern Tiger Swallowtail
mostly black, hindwing with blue scaling and tail; mimics toxic Pipevine Swallowtail

Spicebush Swallowtail
pale green spots on wing margins; hindwing with much greenish-blue scaling; orange eyespot and tail

3–3.5"

1.75–2.5"

White Admiral
broad white spot band across both wings; hindwing with outer red spot band

Baltimore Checkerspot
rows of small white spots; large reddish-orange marginal spots; found in wetlands

3–4"

1.75–2.5"

Mourning Cloak
broad irregular yellow wing borders; outer row of purple-blue spots; often first spring butterfly seen

Red Admiral
forewing with central reddish band, white apical spots; hindwings with broad red border

2.8–3.75"

male

0.9–1.25"

Common Sootywing
shiny black; white spots on
forewing and head

Promethea Silkmoth
brownish black; tan wing borders;
pinkish around dark eyespot

1.8–2.2"

4.9–5.9"

Cecropia Silkmoth
red thorax and forewing bases;
white-striped abdomen; wings have
a crescent spot and a red band with
a white border

Virgin Tiger Moth
forewing black with cream lines;
hindwing and abdomen are pink
to yellow and are dotted with
black spots

1–1.5"

2.4–3.5"

Eight-spotted Forester
forewing with two yellow spots;
hindwing with two white spots;
adults fly during the day

White Underwing Moth
forewing and thorax white
with black circles and spots,
abdomen black; larvae curl up
in a ball when disturbed

1.5–2.0"

Virginia Ctenucha
forewing unmarked dull gray;
hindwing all black; body metallic
blue with orange head; active
during the day; avid flower visitor

Butterflies
of the Northeast

Jaret C. Daniels

Adventure Quick Guides
IDENTIFY BUTTERFLIES WITH EASE

Adventure Quick Guides

Organized by color for quick and easy identification, this guide covers 116 species of the most common butterflies and moths found in the Northeast.

KEY

- If the male and female of a species look the same or nearly the same, only one butterfly is shown.
- When the male and female are different colors, they are shown in their respective color sections with "male" or "female" labels.
- The average size of each species is listed, in inches, to the right of the butterfly.

BUTTERFLY AND MOTH ANATOMY

Knowing the basics about butterfly and moth anatomy is the easiest way to improve your identification skills. The two diagrams below will help you identify the basic parts of a butterfly or moth.

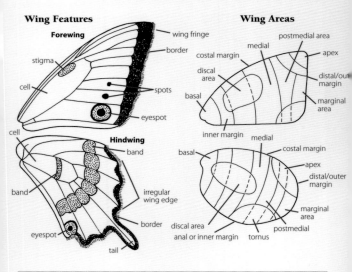

Wing Features

Forewing
- wing fringe
- border
- stigma
- cell
- spots
- eyespot
- cell

Hindwing
- band
- band
- irregular wing edge
- border
- eyespot
- tail

Wing Areas

- medial
- postmedial area
- costal margin
- apex
- discal area
- distal/outer margin
- basal
- marginal area
- inner margin
- medial
- basal
- costal margin
- apex
- distal/outer margin
- marginal area
- discal area
- postmedial
- anal or inner margin
- tornus

10 9 8 7 6 5 4 3

Copyright © 2019 by Jaret C. Daniels
Published by Adventure Publications
An imprint of AdventureKEEN
310 Garfield Street South
Cambridge, Minnesota 55008
(800) 678-7006
www.adventurepublications.net
All rights reserved
Printed in China
ISBN 978-1-59193-826-2 (pbk.)

Cover and book design by Lora Westberg
Edited by Brett Ortler
Cover photo: Great Spangled Fritillary -
Kenneth Keifer/shutterstock.com

male

0.75–1"

Eastern Tailed-Blue
hindwing with orange-capped black spot; the only blue with hindwing tails; males blue, females brownish-gray above

0.75–1"

Spring Azure
color and pattern seasonally variable; sexes differ; dusky blue with black spots and dark scaling along margin; seen in early spring

0.8–1.2"

Summer Azure
chalky blue with faint dark spots

1–1.25"

Silvery Blue
dull gray with band of round white outlined black spots; often local and uncommon

male

1–1.25"

Karner Blue
unmarked bright blue above; whitish gray below with orange submarginal band; endangered; larvae feed on wild lupine

3–4"

Red-spotted Purple
forewing velvety black, hindwing with iridescent blue scaling; mimics toxic pipevine swallowtail; adults feed at tree sap, rotting fruit, and animal dung

4.5–5.5"

Giant Swallowtail

broad crossing yellow spotted bands, hindwing tail with yellow center; adults continuously flutter wings while feeding

1–1.2"

Frosted Elfin

Hindwing with dark irregular central line, silvery scaling toward margin, and a single short tail

0.9–1.25"

Coral Hairstreak

gray-brown, hindwing with coral spot band along margin, tailless; adults fond of milkweed blossoms

1–1.25"

Edwards' Hairstreak

gray-brown wings with white-rimmed black spots

1–1.25"

Banded Hairstreak

band of dark dashes edged outwardly in white; hair-like tails resemble antennae and help deflect attacking predators

1–1.25"

Hickory Hairstreak

dark dashes edged in white on both sides

1–1.3"

Striped Hairstreak

wide, dark bands outlined in white; uncommon and highly localized

0.75–1"

Red-banded Hairstreak

broad red band edged outwardly in white; found in the southern portion of our region

White-M Hairstreak
hindwing with white band forming an "M" shape near red patch

American Snout
forewing with orange basal scaling and apex squared off; gets its name from long snout-like labial palpi; resembles a dead leaf; adults have quick, bouncy flight

Hackberry Emperor
triangular wings, forewing apex black with white spots and a single black eyespot; males regularly fly out to investigate passing organisms; occasionally lands on humans

Tawny Emperor
triangular wings, forewing apex black without white spots

Brown Elfin
wings brown to reddish brown; darker hindwing base with an irregular black line; lacks hindwing tail

Gypsy Moth
female has a tan body and white to cream-colored wings with dark markings; male has brown, mottled wings and ferny antennae

Common Buckeye
conspicuous eyespots, forewing with broad white band; often perches on bare ground; wary and difficult to approach

Dun Skipper
wings mostly unmarked and chocolate brown; underside of hindwing may have faint spot band; upper side of forewing with small white spots in female

Mostly brown

1.75–2.6"

Northern Pearly-eye
violet cast, row of yellow-rimmed dark eyespots; adults often active on cloudy days; doesn't visit flowers

1.6–2.25"

Eyed Brown
light brown; jagged dark line through wings; row of yellow-rimmed dark eyespots

1.1–1.5"

Common Ringlet
forewing with black eyespot near apex, hindwing two-toned with pale wavy central band

1.5–1.8"

Little Wood-Satyr
two dark lines through wings, each wing with two large yellow-rimmed eyespots; adults perch with wings partially open

1.8–2.6"

Common Wood-Nymph
variable eyespot border; may have yellow forewing patch; generally common; open grasslands, prairies, wet meadows; adults visit flowers

1.75–2.5"

Silver-spotted Skipper
elongated forewing, hindwing with large white central patch; largest and showiest skipper in region; larvae construct rolled leaf shelters on host

1.4–1.75"

Hoary Edge
hindwing with hoary white marginal patch

1.2–1.7"

Northern Cloudywing
forewing with small misaligned white spots

1.2–1.6"

Southern Cloudywing
forewing with prominent band of aligned glassy spots

1–1.6"

Dreamy Duskywing
forewing lacks glassy spots

1.5–1.9"

Juvenal's Duskywing
forewing with extensive gray scaling and glassy cell end spot; common to abundant in early spring

1.25–1.75"

Horace's Duskywing
forewing with less extensive gray scaling and usually lacks glassy cell end spot

4.3–5.5"

Ailanthus Silkmoth
Wings brown with pink-edged white central line; each wing with large pale crescent spots

1–1.25"

Peck's Skipper
hindwing with two broad yellow spot bands

0.8–1.25"

Tawny-edged Skipper
forewing with orange scaling along costal margin; hindwing unmarked

female

1–1.4"

Zabulon Skipper
lavender-gray scaling along outer wing margins; hindwing with white bar along apex

0.9–1.2"

Mulberry Wing
wings brownish-black and rounded; underside of hindwing with irregular yellow central patch

female

1–1.5"

Io Moth
forewing reddish-brown, hindwing with large black eyespot and broad pink anal margin

4–5.8"

Polyphemus Moth
hindwing with large oval eyespot and dark postmedian band; males have broad ferny antennae; eyespots may startle predators

female

3–4"

Promethea Silkmoth
wings are two-toned in pink and brown and have dark bases and a pale central line

2.4–3.6"

White-lined Sphinx
forewing with wide diagonal white line from apex; hindwing with pink center; feeds like a hummingbird at flowers; active at dusk and dawn

1.7–3.25"

Twin-spotted Sphinx
forewing with irregular outer margin; hindwing with pink center and large dark eyespot

1.1–1.5"

Northern Broken Dash
hindwing yellow-brown with central band of faint, light spots

1.1–1.5"

Little Glassywing
hindwing purplish-brown with band of faint, light spots

1.5–2.2"

0.9–1.4"

Hummingbird Clearwing

narrow, transparent wings with reddish-brown borders; olive thorax, blackish abdomen; common daytime flower visitor

Eastern Tent Caterpillar Moth

forewing tan to brown with two white central lines; larvae form large silk webs on trees in spring; considered a minor forest pest

1.5–1.75"

Banded Tussock Moth

forewing pale yellow-brown with dark outlined spot bands and pointed apex; hairs on larvae can irritate skin

Acadian Hairstreak
light gray with postmedian band or white-outlined black spots, hindwing with an orange-capped blue patch near tail; adults exceedingly fond of milkweeds

Gray Hairstreak
wings uniformly gray with narrow black line edged in white; hindwing with orange-capped black spot near tail; arguably most common hairstreak in region

Common Checkered-Skipper
black with white spots and dashes; wing fringes checkered

Modest Sphinx
forewing gray-brown with pale gray base; hindwing crimson with dark crescent-shaped eyespot

The Penitent Underwing
forewing gray-brown in a bark-like pattern; hindwing banded orange and black with a pale margin

White-marked Tussock Moth
gray with rounded apex and a pale spot near the anal angle

Pink-spotted Hawkmoth
elongated forewing gray with dark mottling; hindwing gray with black bands and a pink base; elongated, robust abdomen with pink spots

Five-spotted Hawkmoth
Wings with wavy bark-like pattern; abdomen has yellow spots; larvae are called tomato hornworms

Juniper Hairstreak
olive green with white spot band; always found near stands of Eastern Redcedar trees

Luna Moth
pale green, each wing with single eyespot, hindwing with a long, curved tail; often found at lights; long tails help thwart bat attacks

Wavy-lined Emerald
emerald green with thin, pale, wavy lines

Pale Beauty
pale green with two dark outlined pale lines; hindwing margin irregular with stubby tail

1.1–1.3"

Harvester
only butterfly with carnivorous larvae; rarely visits flowers; found in small, localized colonies

1.75–2.75"

Garden Tiger Moth
Forewings brown with irregular white bands; hindwing orange with black-rimmed blue spots

1.3–1.9"

Pearl Crescent
orange with black bands, spots, and borders; widespread and common; larvae feed on asters

0.9–1.4"

American Copper
forewing orange with dark spots and margins; hindwing with wavy orange band

1.25–1.65"

Bronze Copper
forewing orange, hindwing gray with dark spots and an orange marginal band

3.5–4"

Monarch
orange with black veins and borders; forewing with white apical spots; can migrate thousands of miles; larvae found on milkweed

2.9–3.8"

Great Spangled Fritillary
bright orange with black lines and spots; wings dark basally

1.4–2"

Silvery Checkerspot
Tawny-orange with black lines and broad black borders; hindwing has a row of white-centered, somewhat square black spots

2.7–3.3"

Aphrodite Fritillary
bright orange with black lines
and spots; wings dark basally,
forewing with dark basal spot
along inner margin

1.6–2.3"

Silver-bordered Fritillary
orange with black borders
enclosing orange spots;
forewing apex rounded

1.25–1.9"

Meadow Fritillary
forewing elongated with squared
off apex; wings dark basally

2.6–3.3"

Viceroy
orange with black veins and
borders, hindwing with postmedian
black line; resembles monarch;
typically found near wetlands

2.25–3"

Question Mark
forewing apex squared, irregular,
jagged margins; hindwing with
short tail and lavender border;
named for silvery question mark
like spot on hindwing underside

2–2.4"

Eastern Comma
forewing apex squared off, irregular,
jagged margins; hindwing with
stubby tail

1.9–2.4"

Milbert's Tortoiseshell
wings dark with wide yellow
orange band; forewing with
two orange bars, irregular,
jagged margins

1.75–2.4"

Painted Lady
pinkish-orange with dark
marks; forewing apex black
with white spots

1.75–2.4"

American Lady
orange with dark marks and borders; forewing apex squared off and a small white spot in an orange field; common in open, disturbed sites

0.7–1"

Least Skipper
small; rounded wings, forewing dark with an orange border; hindwing an unmarked orange; males have a long pointed abdomen

0.8–1.25"

Arctic Skipper
dark brown dorsally with orange spots; hindwing below yellow-orange with dark-rimmed pale spots

0.9–1.1"

European Skipper
bronzy orange with dark borders, veins darkened toward margins; non-native, from Europe; fond of flowers; common garden butterfly

1.5–1.75"

Leonard's Skipper
hindwing reddish brown with cream-white spot band; adults have strong, powerful flight; occurs in late summer through early fall

male

1.5–1.75"

Sachem
hindwing golden brown with angled pale spot band or patch

1–1.4"

Delaware Skipper
wings unmarked golden orange

1.4–1.6"

Hobomok Skipper
hindwing purplish brown with broad yellow-orange central patch; female has two darker forms

Mostly orange

male

1.4–1.7"

3.9–6"

Zabulon Skipper
hindwing yellow with dark base
enclosing yellow spot; found in
southern portions of the region

Royal Walnut Moth
forewing grayish with orange
veins and yellow spots, hindwing
orange; large, spiny larvae are
called hickory horned devils

Mostly white

2.5–4"

male

1.25–2"

Zebra Swallowtail
white-and-black striped hindwing
with red eyespot and long tail;
found in southern parts of our
region; uncommon in urban areas

Cabbage White
forewing with black tip and single
black eyespot (male); two eyespots
on female; accidentally introduced
from Europe; minor garden pest of
cabbage, broccoli, and cauliflower

1.25–2.2"

Checkered White
white-and-black checkered pattern;
dark scaling seasonally variable

1.25–2"

Orange Sulphur
hindwing with large red-rimmed
silver spot and dark spot band;
common to abundant in clover
and alfalfa fields; males and most
females are bright yellow to
yellow-orange; the above photo
shows the female's white form,
which is uncommon to rare

2.4–3.5"

Giant Leopard Moth
forewing and thorax white with
black circles; black abdomen; largest
tiger moth in eastern U.S.

Mostly yellow

Eastern Tiger Swallowtail

male and female (yellow form) are large with wide black stripes; broad black wing margin, hindwing with a single tail; some females are black in color

3.5–5.5"

Canadian Tiger Swallowtail

large, wide black stripes, broad black wing margin, hindwing with single tail; smaller than Eastern Tiger Swallowtail; females almost always yellow

2.7–3.5"

Clouded Sulphur

yellow to greenish-yellow below, pink wing fringes; hindwing with large red-rimmed silver spot; common to abundant in clover and alfalfa fields

1.9–2.3"

Orange Sulphur

bright yellow, hindwing with large red-rimmed silver spot and dark spot band; common in clover and alfalfa fields; often appears orange in flight; females occasionally white

1.9–2.3"

Pink-edged Sulphur

Plain yellow with pink margins; hindwing with central silver spot ringed in pink

1.5–2.6"

Little Yellow

small, yellow to near white; several dark spots or patches; seasonally colonizes region; low, erratic flight

1–1.6"

male

Io Moth

forewing yellow, hindwing with large dark eyespot and broad pink anal margin; larvae have venomous spines that can inflict a painful sting

2–3.2"

Rosy Maple Moth

forewing with a pink base, outer margin, and a diagonal line from the apex

1.3–2.2"

Mostly yellow

3.4–6.5"

Imperial Moth
forewing yellow with pink-brown mottling; hindwing yellow with broad pink-brown outer margin; larvae do not spin cocoons and pupate underground instead

1.2–2"

Snowberry Clearwing
narrow transparent wings with reddish-brown borders; thorax gold; abdomen black and gold; day-flying

1.1–1.6"

Confused Eusarca
Wings yellow-tan with a dark central line and a small black spot

1.7–2.6"

Isabella Tiger Moth
forewing brownish with pointed apex; hindwing usually paler, often flushed with pink; fuzzy red-and-black larvae

Larvae

1.25–2"

Monarch Caterpillar
body striped with alternating bands of black, white, and yellow; a pair of long, black filaments on both ends

1.75–2.20"

Black Swallowtail Caterpillar
green with black bands and yellow-orange spots

4–4.5"

Cecropia Moth Caterpillar
bluish-green body with bright blue, red, and yellow tubercles (outgrowths), each with short black spines

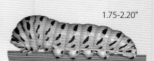

4.5–5.5"

Hickory Horned Devil
greenish-blue body with an orange head; long, curved red-and-black horns

Larvae

2.3–2.9"

2-2.4"

Polyphemus Moth Caterpillar

green with white lateral stripe edged with red; covered in black-tipped green venomous spines; **caution**, spines can cause painful stings

Io Moth Caterpillar

green with white lateral stripe edged with red; covered in black-tipped green venomous spines; **caution**, spines can cause painful stings

1.5-1.7"

2-2.3"

Red-Spotted Purple Caterpillar

mottled green, brown, and cream; two long knobby horns on the thorax; resembles a bird dropping

Tiger Swallowtail Caterpillar

green with enlarged thorax bearing two small false eyes; larva turns brown prior to pupating

2-2.5"

3-4"

Giant Swallowtail Caterpillar

brown with yellow and cream patches; resembles a bird dropping

Tomato Hornworm

bright green body with seven white stripes on the side; a prominent curved horn off the back

3-3.75"

1.8-2.25"

Promethea Silkmoth Larva

stout gray-green body with small black spots surrounded by blue; four red, rounded projections off the front and a single yellow projection off the rear

Banded Wooly Bear

densely fuzzy; banded with black at both ends and reddish brown in the middle

Larvae

2-3"

Luna Moth Larva
green body with small reddish spots and a yellow band

1-1.5"

White-marked Tussock Moth Larva
hairy, gray to cream body with two long black hair tufts, one past the head and one off the rear; four compact cream-colored brushy hair tufts on the back

2.5-3"

Giant Leopard Moth Larva
densely fuzzy; body black and covered with black hairs; distinct red bands around each segment

1.75-2"

Mourning Cloak Larva
black body with a row of crimson patches on the back, fine white speckling and several rows of back, branched spines

JARET C. DANIELS
Jaret C. Daniels, Ph.D., is a professional photographer, author, and entomologist. He has written numerous scientific papers, popular articles, and books on gardening, conservation, insects, and butterflies.

Adventure Quick Guides

Only Northeast Butterflies & Moths
Organized by color for quick and easy identification

Simple and convenient—narrow your choices by color, and view just a few species at a time

- Pocket-size format—easier than laminated foldouts
- Professional photos showing key markings
- Easy-to-use information for even casual observers
- Size ranges for quick comparison and identification
- The basics of butterfly and moth anatomy

Collect all the *Adventure Quick Guides* for the Northeast

ISBN 978-1-59193-826-2 **U.S. $9.95**

5 0 9 9 5

9 781591 938262

PUBLICATIONS
Adventure
an imprint of AdventureKEEN

NATURE/BUTTERFLIES/NORTHEAST